FLORA OF TROPICAL EAST AFRICA

DIPSACACEAE

D. M. NAPPER

Annual or perennial herbs or subshrubs. Leaves opposite, entire, lobed or variously dissected, often markedly heterophyllous. Inflorescences of one or several long-pedunculate dense heads each surrounded by an involucre. Receptacle convex, globose or cylindric, naked, shortly pilose or with paleaceous bracts each subtending a single flower. Flowers sessile, bisexual (excepting sometimes the marginal ones), with the ovary surrounded by and enclosed in a highly characteristic tubular involucel formed by the fusion of 2 or 4 bracts often with a membranous cupular or spreading lobed limb. Calyx cupular or saucer-shaped with an entire, irregularly 4–5-lobed or toothed margin, sometimes with long naked or plumose bristles. Corolla gamopetalous with an irregularly 4- or 5-lobed limb and spreading lobes. Stamens 2–4, alternating with the corolla-lobes and inserted near the top of the tube; anthers dithecous, oblong, exserted. Ovary inferior and unilocular with a solitary ovule; style filiform with a terminal or lateral fusiform oblique or bifid stigma. Fruit an achene closely surrounded by the persistent involucel and crowned by the persistent calyx.

A small family of about 8 genera, occurring in the temperate regions of Europe and Asia, especially abundant in the Mediterranean region and the Near East, but also well represented in southern Africa, less common in tropical Africa.

It is of little economic importance apart from the Fullers' Teasel, *Dipsacus fullonum* L. subsp. *fullonum*, used for carding wool.

Calyx without conspicuous bristles:
    Involucral bracts narrowly lanceolate, leafy; stem
        aculeate . . . . **1. Dipsacus**
    Involucral bracts small, ovate to ovate-lanceolate, of
        similar texture to the bracts of the receptacle;
        plants not prickly . . . . . . **2. Cephalaria**
Calyx with bristles at least 2–3 mm. long:
    Calyx bristles naked, often scabrid; receptacle bracts
        present, linear-lanceolate . . . . . **3. Scabiosa**
    Calyx bristles plumose; receptacle bracts absent . **4. Pterocephalus**

1. DIPSACUS

L., Sp. Pl.: 97 (1753) & Gen. Pl., ed. 5: 43 (1754)

Erect perennial herbs with aculeate, or rarely smooth, stems. Leaves pinnatipartite or entire, sometimes the stem-leaves connate at the base. Inflorescences terminal or terminal and lateral. Heads involucrate with 1 or 2 rows of foliaceous often spine-tipped bracts of variable length. Bracts of the receptacle acute, chartaceous and spine-tipped, usually shorter than the involucral bracts. Involucel with or without a narrow cupular limb and developing 8 longitudinal furrows. Calyx cupular, small, undulate or minutely toothed, usually pubescent or pilose. Corolla 4-fid, unequally lobed. Stamens 4, exserted. Stigma entire, rarely shortly bifid.

A genus of about 15 species occurring chiefly in Europe, Asia and northern India. Two or three species are reported from tropical Africa, occurring most abundantly in the north-east, but reaching to the Cameroon Mt. in the west and the Crater Highlands of Tanganyika in the south.

D. pinnatifidus *A. Rich.*, Tent. Fl. Abyss. 1: 367 (1847); Hiern in F.T.A. 3: 250 (1877); Engl., Hochgeb. Trop. Afr.: 403 (1892); P.O.A. C: 396 (1895); Z.A.E.: 342 (1912); De Wild., Pl. Bequaert. 1: 555 (1922); F.P.N.A. 2: 385 (1947); F.P.S. 2: 467 (1952); A.V.P.: 183 (1957); F.W.T.A., ed. 2, 2: 223 (1963); E.P.A.: 1028 (1965). Type: Ethiopia, Begemedir, *Schimper* 665 (BM, K, iso.!)

Perennial herbs 0·6–3 m. high. Stems erect, usually stout, angular and bearing small prickles on the angles, rarely terete and smooth, glabrous throughout or with the lower internodes pubescent. Leaves sessile, 5–20 cm. long, lanceolate, rounded below, toothed or pinnatipartite, with the midrib and sometimes the secondary veins aculeate beneath, usually glabrous; bases of the upper stem-leaves sometimes connate; lateral shoots with much smaller and less divided subpetiolate or sessile leaves. Terminal inflorescences of one to several heads on aculeate peduncles resembling the stems, often more densely aculeate and pubescent beneath the globose 2·5–4(–4·5) cm. diameter head; lateral inflorescences frequently developed on older plants, more contracted, with heads 2–3 cm. in diameter. Involucral bracts spreading, foliaceous, attenuate, from half the length of to slightly exceeding the head. Bracts of the receptacle chartaceous, acuminate, spine-tipped and very variable in size, especially in the length of the pubescent spiny tip. Involucel pubescent, becoming 3–4 mm. long in fruit with a narrow sinuate membranous limb when mature. Calyx cupular, 1–1·5 mm. long, shallowly 4-lobed, pilose. Corolla white or cream, 6–15 mm. long; tube sparingly pubescent without; lobes glabrous, subequal. Fig. 1.

UGANDA. Karamoja District: Mt. Moroto, Jan. 1959, *J. Wilson* 644!; Kigezi District: Behungi, 23 Dec. 1933, *A. S. Thomas* 1202!; Mbale District: Elgon, Apr. 1930, *Liebenberg* 1614!
KENYA. Northern Frontier Province: Mt. Nyiru, July 1960, *Kerfoot* 2003!; Elgon, 21 May 1948, *Hedberg* 1027!; Mt. Kenya, 26 Dec. 1934, *Gedye* 64 in *C.M.* 6812!; Masai District: Mau area, Olokurto, 1 June 1961, *Glover et al.* 1519!
TANGANYIKA. Masai District: Ololmoti Volcano, 16 Sept. 1932, *B. D. Burtt* 4354! & W. side of Mt. Oldeani, 16 Feb. 1961, *Newbould* 5705!
DISTR. U1–3; K1, 3, 4, 6; T2; also from the Cameroon Mts. to Ethiopia and the Congo Republic
HAB. Upland grassland and damp places, glades in upland forest and bamboo thicket; 2000–3950 m.

SYN. *D. appendiculatus* A. Rich., Tent. Fl. Abyss. 1: 367 (1847). Type: Ethiopia, Begemedir, *Schimper* 856 (K, iso.!)
    D. schimperi A. Br. in Schweinf. Beitr.: 287 (1867), *nomen* & in App. ad Ind. Sem. H. Berol. 1867: 5 (1867). Type: a plant cultivated in the Berlin Botanic Gardens
    D. pinnatifidus A. Rich. var. *integrifolius* Engl. in E.J. 19, Beibl. 47: 49 (1894); P.O.A. C: 396 (1895); F.P.N.A. 2: 385 (1947). Types: Tanganyika, Kilimanjaro, *Volkens* 862 & 967 (BM, isosyn.!) & 1550 (E, isosyn.!)
    D. bequaertii De Wild., Pl. Bequaert. 1: 554 (1922); F.P.N.A. 2: 386 (1947). Type: Congo Republic, Ruwenzori Mts., *Bequaert* 4655 (BR, holo.!, EA, iso.!)
    D. kigesiensis R. Good in J.B. 62: 334 (1924). Type: Uganda, Kigezi, *Misses Godman* 189 (BM, holo.!)

VARIATION. The considerable degree of vegetative variation gives rise to certain forms which, at their extreme development, are readily separable but which are linked by such a close gradation that no infraspecific taxa may be delimited at present. Two forms are particularly distinct: 1, stems very robust and stiffly erect, with large entire leaves with at least the upper connate at the base; receptacle bracts very long acuminate: 2, stems more slender, occasionally taller but often weak and straggly; leaves not connate, often smaller and deeply divided; receptacle bracts acuminate or acute. The former appears to occur most frequently in exposed sites, the latter more often in forest glades, but further ecological and climatological data are required.

D.E.

FIG. 1. *DIPSACUS PINNATIFIDUS*—**1**, habit, × ½; **2**, head, showing bract variation, × ⅔; **3**, involucel and flower, × 6; **4**, fruit with persistent calyx, × 6. 1, from *Napier* 664; 2, 3, from *Mooney* 8305; 4, from *Gillett* 14703.

2. CEPHALARIA

Roem. & Schult., Syst. Veg. 3: 1 (1818) & Mant.: 2 (1827); Szabó in Mat. Term. Kozlem. 38, 4: 1–248 (1940), *nom. conserv.*

Perennial or annual herbs or subshrubs (not in Africa). Leaves very variable, most species heterophyllous. Inflorescence terminal. Heads involucrate with short obtuse oblanceolate to ovate scarious bracts in one or two rows. Receptacle bracts present, scarious, longer than those of the involucre, obtuse, acute or acuminate and frequently with a pungent tip. Involucel usually smooth when immature becoming 8-furrowed in fruit, glabrous or pilose, with a small 4–8-toothed or -lobed limb, more rarely the limb entire or crenate. Calyx small, spreading, 4–many-lobed, pilose or villous. Corolla campanulate, 4-fid, unequally lobed, larger in the marginal than in the inner flowers. Stamens 4. Style filiform with an entire oblique stigma.

A genus of about 60 species occurring in southern Europe, the Near East and South Africa. A few species occur in tropical Africa.

Involucels glabrous or pubescent with short lobes:
  Stems with short leafy shoots from nearly every node;
    leaves numerous and all alike .   .   .   . 1. **C. goetzei**
  Stems unbranched below the inflorescence; stem-
    leaves smaller and of different shape from the
    radical ones .   .   .   .   .   . 2. **C. pungens**
Involucels villous, long-toothed:
  Bracts acuminate, pungent; all leaves petiolate . 3. **C. integrifolia**
  Bracts rounded, obtuse; stem-leaves sessile, broadly
    linear .   .   .   . 4. **C. katangensis**

1. **C. goetzei** *Engl.* in E.J. 30: 418 (1902); Szabó in Magyar Bot. Lap. 24: 14 (1926) & in Mat. Term. Kozlem. 38, 4: 130 (1940). Type: Tanganyika, Mbeya District, Umalila, *Goetze* 1349 (BM, iso.!)

Perennial herb up to 60 cm. high. Stems erect, terete or angular, glabrous or basal internodes sparingly pilose with a sparse retrorse indumentum. Basal leaves very varied, narrowly elliptic, entire or lanceolate, coarsely toothed, lobed or pinnatipartite, pubescent or glabrous; stem-leaves sessile, linear or narrowly elliptic and cuneate below, with or without 1 or 2 pairs of small linear lobes towards the base, or pinnatipartite with entire linear lobes. Head subglobose, 2·5–3 cm. in diameter. Involucral bracts ovate. Receptacle bracts oblanceolate, acuminate, excepting the lowest ones which are usually intermediate with the involucral bracts, silvery pubescent with dark tips. Involucel subglabrous with 4 broad obtuse lobes. Calyx villous. Corolla white, 10–15 mm. long. Fig. 2/9.

TANGANYIKA. Dodoma District: Kilimatinde, 1 Feb. 1904, *Prittwitz* 102!; Mbeya District: N. Usafwa Forest Reserve, Sept. 1959, *Procter* 1363!; Songea District: 11 km. W. of Songea, 1 Jan. 1956, *Milne-Redhead & Taylor* 8019!
DISTR. T5, 7, 8; Malawi
HAB. Upland and boggy grassland, large open spaces in woodland; 960–2250 m.

2. **C. pungens** *Szabó* in E.J. 57: 642 (1922) & in Magyar Bot. Lap. 24: 14 (1926). Type: Tanganyika, Kipengere Mts., *Goetze* 979 (G, iso.!)

Perennial herb up to 1 m. high, rarely up to 1·5 m. Stems erect, unbranched, terete or 6-sided, glabrous, rarely the lower internodes with scattered retrorse hairs. Leaves very variable but the radical ones narrowly lanceolate to broadly elliptic, cuneate below, more rarely elongate with a very

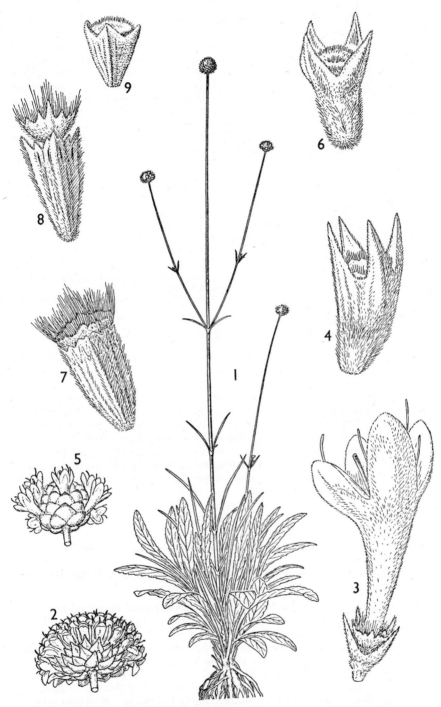

FIG. 2. *CEPHALARIA INTEGRIFOLIA*—**1**, habit, × ⅓; **2**, head, × 1; **3**, involucel and flower, × 6; **4**, fruit, × 6. *C. KATANGENSIS*—**5**, head, × 1; **6**, involucel, × 6. *C. PUNGENS*—**7**, involucel, × 6; **8**, same, showing variation, × 6. *C. GOETZEI*—**9**, involucel, × 6. 1, 2, 4, from *Milne-Redhead & Taylor* 10828; 3, from *Richards* 12971; 5, from *Quarré* 3146; 6, from *Richards* 1364; 7, from *Watermeyer* 14; 8, from *Mortimer* T 228; 9, from *R. M. Davies* 668.

long petiole, 8–20 cm. long, glabrous or sparingly pilose, sometimes with bulbous-based hairs; lower stem-leaves similar, narrowly elliptic or lanceolate, rarely lobed at the base, 5–25 cm. long, 1–2·3 cm. wide, glabrous or sparingly pilose with ciliate margins; upper stem-leaves linear with 1–2 pairs of linear lobes towards the base, more rarely toothed or entire. Heads subglobose, 2–4 cm. in diameter. Involucral bracts ovate, pubescent. Receptacle bracts lanceolate, cuspidate, dark grey, pilose and ciliate or with a short dense silvery pubescence, the lower ones intermediate to the involucral bracts. Involucel shortly pubescent, having a limb with 4 entire or tridentate lobes or entire, glabrous and membranous. Calyx villous, deeply fimbriate. Corolla white, the marginal flowers usually larger than the inner. Fig. 2/7, 8, p. 5.

KENYA. W. Suk District: Sondang Hills, Jan. 1935, *Thorold* 2773!; Baringo District: Kamasia Hills, Sept. 1938, *Gardner* in *F.D.* 1238!; Nakuru District: Mau Forest, Bondui, 12 Jan. 1946, *Bally* 4782!
TANGANYIKA. Mbeya Mt., 13 May 1956, *Milne-Redhead & Taylor* 10224!; Iringa District: Mufindi, Ngowasi Lake, 24 Mar. 1962, *Polhill & Paulo* 1858!; Njombe District: Elton Plateau, 12 Nov. 1931, *R. M. Davies* E. 25!
DISTR. **K**2, 3; **T**7; throughout southern tropical Africa and to Swaziland
HAB. Upland grassland often in damp places and in large forest glades; 1800–3000 m.

SYN. [*C. centauroïdes* sensu Engl., Hochgeb. Trop. Afr.: 404 (1892) & P.O.A. C: 395 (1895), pro parte, *non* (Lam.) Roem. & Schult.]
[*C. attenuata* sensu Engl. in E.J. 30: 418 (1902), *non* (Thunb.) Roem. & Schult.]

3. C. integrifolia *Napper* in K.B. 21: 463 (1968). Type: Tanganyika, Songea, *Milne-Redhead & Taylor* 10828 (K, holo.!, EA, iso.!)

Perennial herb up to 1·8 m. high. Stems erect, unbranched below the inflorescence, terete or 6-sided, glabrous, rarely the lower internodes with scattered retrorse hairs. Radical and lower stem-leaves elongate lanceolate, long-cuneate below, entire or toothed, glabrous, rarely sparingly setose; stem-leaves entire or more rarely pinnatipartite and smaller. Heads hemispherical, 2·5–3·5 cm. in diameter. Involucral bracts blackish, ovate-lanceolate. Receptacle bracts acuminate or cuspidate often with a long spinous tip, grey tomentose with villous margins or shortly silvery pubescent with darker tips. Involucel with 4 long teeth, densely villous. Calyx small, villous. Corolla 8–10 mm. long, white or cream, marginal flowers longer than the inner. Fig. 2/1–4, p. 5.

TANGANYIKA. Ufipa District: Nsangu Forest, 6 Aug. 1960, *Richards* 12971!; Iringa District: Iheme, *McGregor* 65!; Songea District: 8 km. W. of Songea, 18 June 1956, *Milne-Redhead & Taylor* 10828!
DISTR. **T**4, 7, 8; Malawi
HAB. Damp ironstone outcrops, seasonally swampy or upland grassland; 990–2700 m.

4. C. katangensis *Napper* in K.B. 21: 464 (1968). Type: Congo Republic, Katanga, *Verdick* 577 (BR, holo.!)

Perennial herb up to 2 m. high. Stems erect, rarely branched below the inflorescence, 6-sided, glabrous, but sometimes sparingly pilose below. Radical leaves elliptic to linear-lanceolate, rounded or acute above, cuneate below with the margins entire, serrate in the upper part only, coarsely toothed or even lobed below, up to 50 cm. or more long, glabrous or sparingly ciliate; stem-leaves linear-lanceolate with a few entire lobes below and scattered teeth in the upper part. Heads dense, hemispherical, 2–2·5 cm. in diameter. Involucral bracts broadly ovate, obtuse, finely pubescent. Receptacle bracts elliptic-ovate, obtuse to rounded, 5–6 mm. long, shortly silvery pubescent with darker tips. Involucel densely pilose, with 4 acute teeth. Calyx small, densely villous. Corolla white, up to 12 mm. long, the marginal flowers larger than the inner. Fig. 2/5, 6, p. 5.

TANGANYIKA. Mpanda District: Kabungu, Feb. 1953, *Friend* 346!
DISTR. **T4**; Zambia and Congo Republic (Katanga)
HAB. *Brachystegia* woodland; 950–1150 m.

SYN. *C. attenuata* (Thunb.) Roem. & Schult. var. *longifolia* De Wild. in Ann. Mus.
Congo, Bot., sér. 4, 3: 164 (1903); Th. & H. Dur., Syll. Fl. Cong.: 288 (1909);
De Wild., Contrib. Fl. Kat.: 218 (1921). Type: Congo Republic, Katanga,
Verdick 577 (BR, holo.!)
[*C. humilis* sensu Szabó in Magyar Bot. Lap. 24: 4 (1926), pro parte; De Wild.,
Contrib. Fl. Kat., suppl. 1: 95 (1927), *non* (Thunb.) Roem. & Schult.]

3. SCABIOSA

L., Sp. Pl.: 98 (1753) & Gen. Pl., ed. 5: 43 (1754)

Erect perennial or annual herbs, usually markedly heterophyllous. Leaves
entire or deeply divided. Inflorescence terminal. Heads involucrate with 1–2
rows of foliaceous bracts shorter or longer than the flowers. Receptacle bracts
linear-lanceolate, much shorter than the flowers, glabrous or pubescent.
Involucel smooth or 8-furrowed towards the top only or throughout its
length, with a membranous many-veined entire or subentire erect or spreading limb. Calyx small with 5 teeth each produced into a long scabrid or
barbed bristle; bristles subequal, spreading in fruit. Corolla 5-fid (4–6-fid
flowers occur sometimes in the same head); lobes unequal. Marginal flowers
are usually considerably longer than the inner ones with a more asymmetric
corolla. Stamens 4. Stigma oblique, entire.

A genus of about 100 species, widespread through the temperate regions of Europe
and Asia, and also occurring in tropical and southern Africa.

For many years the two most widely distributed African forms have been regarded as
mere colour variants of one species, *S. columbaria* L. Their differences, however, though
slight and tending to be obscured by the wide range of variation exhibited by each, are
sufficient to warrant separation, more particularly as the affinities of the one appear
to lie with the Mediterranean and temperate European flora and of the other with the
South African flora.

Corolla lilac or deep mauve; calyx bristles 3–5·5 mm.
  long; median stem-leaves subentire or with a large
  terminal lobe, rarely ± equally divided . . **1. S. columbaria**
Corolla white or cream (rarely pinkish); calyx bristles
  5·5–10 mm. long; median leaves ± equally lobed **2. S. austro-africana**

1. S. columbaria *L.*, Sp. Pl.: 99 (1753); Coulter, Mem. Dips.: 38 (1823)
& in DC. Prodr. 4: 658 (1830); A. Rich., Tent. Fl. Abyss. 1: 368 (1847);
Hiern in F.T.A. 3: 252 (1877); Engl., Hochgeb. Trop. Afr.: 404 (1892);
P.O.A. C: 396 (1895); Hiern, Cat. Welw. Afr. Pl. 1: 518 (1898); A.V.P.:
184 (1957); E.P.A.: 1030 (1965). Type: Sweden, Gotland, *Linnaeus* (LINN,
holo., K, photo.!)

Perennial herb, 0·15–1·2 m. high, with erect branched pubescent stems,
markedly heterophyllous. Radical leaves sessile, entire or deeply lobed in the
lower part with the margin toothed or crenate above, glabrous or pubescent;
lower stem-leaves deeply pinnatifid, rarely bipinnate; upper stem-leaves
deeply pinnatipartite, smaller and simply pinnate or entire. Heads (2–)2·5–
3·5(–4) cm. in diameter, on long peduncles. Involucral bracts lanceolate,
acute, varying from half the length of the marginal flowers to exceeding them.
Bracts of the receptacle linear-lanceolate, acute, 3–5 mm. long, pubescent
above. Involucel pubescent, becoming 8-furrowed and (including the limb)
4–5 mm. long in fruit. Calyx small, pilose, with bristles 3–6 mm. long.
Corolla lilac, mauve or, rarely, pinkish, 5–20 mm. long, larger and more
unequally lobed in the marginal flowers. Fig. 3/5, p. 8.

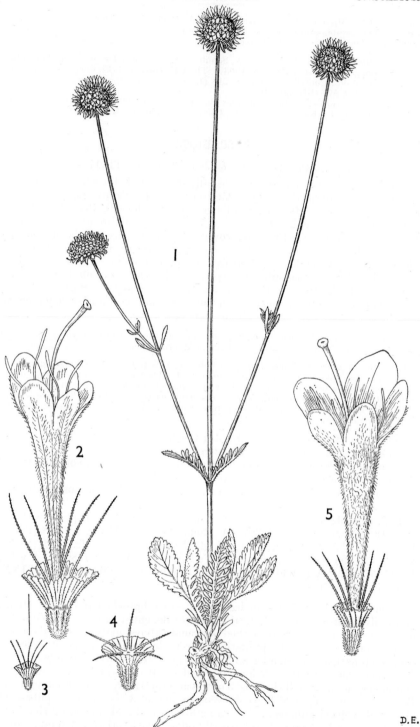

D.E.

Fig. 3. *SCABIOSA AUSTRO-AFRICANA*—**1**, habit, × ½; **2**, involucel and flower, × 6; **3**, same, with corolla removed, drawn to scale of 4, × 2; **4**, fruit with calyx, × 2. *S. COLUMBARIA*—**5**, involucel and flower, × 6. 1, 4, from *Bullock* 1415; 2, 3, from *Haarer* 425; 5, from *Maas Geesteranus* 5494.

UGANDA. Karamoja District: Mt. Moroto, Feb. 1936, *Eggeling* 2858!; Elgon, Apr. 1930, *Liebenberg* 1627! & 12 Nov. 1933, *Tothill* 2404!

KENYA. Northern Frontier Province: Mt. Nyiru, 30 Dec. 1955, *J. Adamson* 555!; Kisumu-Londiani District: Tinderet Forest Reserve, 14 July 1949, *Maas Geesteranus* 5494!; Aberdare Mts., Sattima, 13 Mar. 1922, *Fries* 2349!

TANGANYIKA. Mbulu District: Oldeani Mt., Jan. 1935, *Moreau* 38! & Mt. Hanang, 26 Dec. 1929, *B. D. Burtt* 2266!; Kilimanjaro, 22 Feb. 1934, *Greenway* 3747!

DISTR. U1, 3; K1, 3–5; T2; southern Africa, Ethiopia, Europe and the Near East

HAB. Upland grassland and grass-moor; 2100–4100 m.

SYN. *S. columbaria* L. var. *robusta* Engl., P.O.A. C: 396 (1895), *nomen nudum*
S. columbaria L. var. *angusticuneata* Engl. in E.J. 19, Beibl. 47: 49 (1894). Types: Tanganyika, Kilimanjaro, *Volkens* 918 & 1199 (both K, isosyn.!)

NOTE. Despite the high degree of variation to be observed in the East African material the gradation of forms is so intricate that separation into infraspecific taxa has proved impossible without further investigations into ecological and other fields such as are beyond the scope of this Flora.

2. **S. austro-africana** *Heine* in Mitt. Bot. Staats. München 1: 445 (1954). Type: South Africa, Cape Province, *Ecklon* 725 (M, holo.!, K, iso.!)

Perennial herb, 0·15–1·2 m. high, with erect pubescent or glabrous stems. Leaves markedly heterophyllous, glabrous or pubescent, the radical ones sessile, toothed or lobed in the lower part, lanceolate, often narrowing to a scarcely winged midrib and appearing lanceolate; lower stem-leaves similar to the radical ones but more deeply lobed, 2–24 cm. long, 0·5–3 cm. broad, with the lobes narrowly linear, rarely toothed or entire; upper stem-leaves smaller, deeply dissected with linear segments. Inflorescence of 1 to several heads (2–)2·5–3·5(–4) cm. in diameter. Involucral bracts lanceolate, pubescent, from half the length of the marginal flowers to longer than them. Bracts of the receptacle linear-lanceolate, acute, 3–5 mm. long, pubescent above. Involucel 8-furrowed, pubescent, becoming 4–6 mm. long in fruit including the membranous, entire, 1·5–2 mm. long, conspicuously veined limb. Calyx small, pilose, with bristles 5·5–10 mm. long. Corolla white, rarely pinkish or lightly tinged with mauve, 0·5–2 cm. long, larger and unequally lobed in the marginal flowers, smaller and more equally lobed in the inner flowers. Fig. 3/1–4.

KENYA. Machakos/Masai District: Chyulu Hills, 21 Apr. 1938, *Bally* 7995!

TANGANYIKA. W. Usambara Mts., Mtai–Malindi road, 19 May 1953, *Drummond & Hemsley* 2667!; Kondoa District: Bereku Ridge, 17 Jan. 1928, *B. D. Burtt* 1138!; Mbeya Airport, 8 Apr. 1956, *Semsei* 2429!

DISTR. K4/6; T1–8; Rwanda Republic, Congo Republic, and throughout southern Africa

HAB. Upland grassland and open bushland, often in places subject to regular burning; 1200–1900(–2500) m.

SYN. [*S. columbaria* sensu Hiern in F.T.A. 3: 252 (1877), pro parte; Engl., Hochgeb. Trop. Afr.: 404 (1892), pro parte, *non* L.]

4. PTEROCEPHALUS

Adans., Fam. 2: 152 (1763)

Perennial or annual herbs or subshrubs. Inflorescence a solitary involucrate head, with 1–2 series of unequal foliaceous bracts shorter than the flowers or exceeding them. Receptacle shortly conical, without bracts. Involucel 8-furrowed, obscurely or irregularly toothed or with a narrow undulate entire membranous limb. Calyx spreading, small, with numerous long plumose bristles of equal length. Corolla 5-fid (4–6-fid flowers sometimes occur in the same head) with unequal lobes, the marginal flowers usually larger and with a more unequally lobed limb. Stamens 4. Stigma oblique, entire.

FIG. 4. *PTEROCEPHALUS FRUTESCENS*—**1,** habit, × ⅔; **2,** involucel and flower, × 3; **3,** fruit with calyx, × 3. 1, 3, from *Greenway* 9592; 2, from *Eichinger* 3467.

A genus of about 25 species, centred in the Middle East, but also occurring in southern Europe and India, with one species in NE. Africa reaching its southern limit in Tanganyika.

P. frutescens *A. Rich.*, Tent. Fl. Abyss. 1: 369 (1847); E.P.A.: 1030 (1965). Type: Ethiopia, Tigre, *Schimper* 235 (K, iso.!)

Perennial herb or subshrub, 20–60 cm. high, with erect or decumbent pubescent stems becoming woody at the base. Leaves sessile, linear-lanceolate to elliptic-lanceolate, mucronate, entire or with a few short linear lobes, 1–5 cm. long, 2–8 mm. wide, sparingly pubescent or glabrous; basal leaves not differentiated. Heads solitary, 2·5–4·5 cm. in diameter. Involucral bracts subbiseriate, lanceolate, acute, up to 2 cm. long, 4 mm. wide, pubescent. Involucel up to 6 mm. long in fruit, with a subentire or irregularly toothed margin, densely pilose or villous. Calyx small, flat, with plumose bristles ± as long as the fully expanded corolla-tube, up to 13 mm. long in fruit. Corolla deep pink, rarely mauvish or very pale, pubescent on the outside, very variable in size with the marginal flowers up to 20 mm. long, the inner flowers smaller. Fig. 4.

KENYA. Naivasha District: Mt. Longonot, 29 Mar. 1930, *Napier* 204!; Meru District: near Meru, 13 Feb. 1922, *Fries* 1517!; Masai District: S. end of Chyulu Hills, 23 June 1953, *Bally* 8983!
TANGANYIKA. Masai District: Ngorongoro Crater, 23 Nov. 1956, *Greenway* 9052!; Mbulu District: Mt. Hanang, 2 Sept. 1932, *B. D. Burtt* 4026!; Arusha District: E. Mt. Meru, Usa R. to Engare Nanyuki, 27 Oct. 1959, *Greenway* 9592!
DISTR. **K3**, 4, 6; **T2**; Ethiopia and the Somali Republic
HAB. Dry rocky hillsides, upland grassland; 1500–2800 m.

SYN. *Scabiosa frutescens* (A. Rich.) Hiern in F.T.A. 3: 252 (1877); Engl., Hochgeb. Trop. Afr.: 405 (1892)

INDEX TO DIPSACACEAE